Where Are t

Dona Herweck Rice

Publishing Credits

Rachelle Cracchiolo, M.S.Ed., *Publisher*
Conni Medina, M.A.Ed., *Managing Editor*
Nika Fabienke, Ed.D., *Content Director*
Véronique Bos, *Creative Director*
Shaun N. Bernadou, *Art Director*
Seth Rogers, *Editor*
John Leach, *Assistant Editor*
Courtney Roberson, *Senior Graphic Designer*

Image Credits: All images from iStock and/or Shutterstock.

Library of Congress Cataloging-in-Publication Data

Names: Rice, Dona, author.
Title: Where are the animals? / Dona Herweck Rice.
Description: Huntington Beach, CA : Teacher Created Materials, [2019] | Audience: K to Grade 3. |
Identifiers: LCCN 2018029756 (print) | LCCN 2018031454 (ebook) | ISBN 9781493899456 () | ISBN 9781493898718
Subjects: LCSH: Animal behavior--Juvenile literature.
Classification: LCC QL751.5 (ebook) | LCC QL751.5 .R56 2019 (print) | DDC 591.5--dc23
LC record available at https://lccn.loc.gov/2018029756

Teacher Created Materials
5301 Oceanus Drive
Huntington Beach, CA 92649-1030
www.tcmpub.com

ISBN 978-1-4938-9871-8

Printed in China
Nordica.082018.CA21800936

The were

pigs

in the  .

sand

Did they all come

? No.

home

The were

ducks

in the .

pond

Did they all come

? No.

The goats were

goats

in the  .

tree

Did they all come

? No.

home

The cows were

cows

in the  .

snow

Did they all come

? No.

home

The  were

dogs

in the .

car

Did they all come

? No.

home

High-Frequency Words

New Words

all
come
did
no
were

Review Words

in
the
they